Rancho La Brea

Orlando   Austin   New York   San Diego   Toronto   London

Visit *The Learning Site!*
www.harcourtschool.com

The Tar Ranch

Prehistoric animals? Caught in tar? In Los Angeles? Thousands of them?

You can see the remains of these animals at Rancho La Brea, located in the 5800 block of Wilshire Boulevard in downtown Los Angeles, California. Rancho La Brea is Spanish for "the tar ranch." At Rancho La Brea, crude oil has been seeping up out of the ground for about 40,000 years. The less dense parts of the oil evaporate, and heavy, black, sticky, tarlike asphalt is left behind. For about the past 40,000 years, animals have been getting caught in the sticky asphalt. At Rancho La Brea, scientists have found remains from more than 465 different kinds of animals, including saber-toothed cats, dire wolves, mammoths, and mastodons, as well as camels, ground sloths, and a species of antelope that stood only 60 centimeters (2 ft) high!

The oil at Rancho La Brea comes from a deposit below Earth's surface called the Salt Lake Oil Field. This oil deposit formed from tiny sea creatures called plankton millions of years ago when this part of California was covered by the Pacific Ocean. Humans arrived

Thousands of animals have been caught in the tarlike asphalt at Rancho La Brea.

in the area about 10,000 years ago and began putting the asphalt to use. Local Native Americans put the sticky asphalt on their canoes and baskets to make them waterproof. Later Spanish settlers used the asphalt to make the roofs of their houses waterproof.

In 1828, a Mexican land grant gave the land with the asphalt pools to Antonio Jose Rocha. At that time, Mexico still ruled California. The grant called the land "Rancho La Brea." Señor Rocha used the land as a cattle ranch. However, his land grant required him to let people continue to take asphalt from the pools when they needed it.

Major Henry Hancock, a lawyer from New Hampshire, bought the 1,720-hectare (4,400-acre) Rancho La Brea from Señor Rocha's son Jose in 1860. Major Hancock made a business out of mining and selling the asphalt. People from as far away as San Francisco bought Hancock's asphalt to use on their roofs. At that time, Rancho La Brea was about 11 kilometers (7 mi) west of the town of Los Angeles. Since then Los Angeles has grown, and now Rancho La Brea, asphalt pools and all, is in the heart of the city.

Rancho La Brea around 1900

Caught!

When he owned the ranch, Señor Rocha thought the bones that people found in the asphalt pools were from his cattle. It wasn't until 1875 that anyone realized that these bones were from other kinds of animals. And it wasn't until 1906 that scientific excavation of the pools began. Since then, nearly 3.5 million fossils have been collected at Rancho La Brea.

The animal fossils at Rancho La Brea formed when the animals became trapped in the asphalt. Some animals probably became trapped while looking for food or water. Others may have been chased into the pools by predators. When they were unable to move, some of the trapped animals were eaten. Others slowly sank into the asphalt and became fossils. The soft parts of the animals decayed, but the hard bones and teeth remained.

The asphalt at Rancho La Brea preserved not only large animal bones but also plants, insects, and other microfossils, some of which are too small to be seen with the naked eye. In fact, the deposits at Rancho La Brea preserved samples from the entire ecosystem. There are fossils from three species of fish, five kinds of amphibians, one type of turtle, and five kinds of snakes. More than one million fossils come from invertebrates. And hundreds of thousands come from plants.

All of these fossils make it possible for scientists to tell what kind of environment existed in Los Angeles 12,000 to 40,000 years ago. The plant fossils at Rancho La Brea provide a continuous record of the climate. What they tell us is that 40,000 years ago the area was a bit cooler and more moist than it is today. This was the end of the Ice Age, but there was no ice here. Instead, the climate was about like that of the Monterey Peninsula, which is 480 kilometers (300 mi) north of La Brea.

Most of the time, the asphalt was hard enough for most animals to cross without problem. However, problems developed when the temperature was particularly warm, which usually happened in the

Camels and dwarf pronghorns are two of the unusual animals that got caught at Rancho La Brea. It's hard to believe, but camels originated in North America. Two to three million years ago, some of them migrated to Eurasia and Africa.

summer. Then, warm temperatures turned the asphalt into a gooey liquid. Fossil evidence indicates shallow ponds and marshes existed at La Brea as well as intermittent streams. Sometimes, water pooled on top of the asphalt to give it the appearance of a lake. At other times, leaves covered the surface and hid the pools. The water would lure animals, and the leaves would camouflage the asphalt. Still, the fossils found so far indicate that relatively few animals got caught—perhaps 10 animals every 30 years.

Scientists noticed one thing that was very odd. In any given area, there are usually many more prey animals than predators. But at Rancho La Brea, scientists found many more fossils from meat-eating predators than from plant-eating prey. Scientists believe that when one large prey animal got caught, it would attract many predators that would also get stuck in the asphalt.

The fossils show that many of the animals trapped at Rancho La Brea were the same species as the animals that live around Los Angeles today. These include bobcats, coyotes, gray foxes, pumas, jaguars, weasels, skunks, and raccoons. Of the one million invertebrate fossils found so far, only two extinct species—two types of dung beetles—are represented. All other invertebrate species are still around. Most of the large mammals, however, are either completely extinct or are not found anywhere near Los Angeles. In addition to those already mentioned, these include the western horse, tapirs, llamas, peccaries, and the American lion, which may have been the same species as today's African lion.

Teratorns

Fossils from more than two dozen species of birds have been recovered at Rancho La Brea. Some of these birds, including the bald eagle and the golden eagle, still live in and around California. Others, such as the teratorns, are extinct. Teratorns were similar in some ways to condors. They had narrow, hooked beaks that were used to pull apart prey. Some fed on dead animals, as condors do today. Other teratorns probably went after their prey by following it on the ground.

A Merriam's teratorn is one bird species that was preserved at Rancho La Brea when it became trapped in the asphalt, probably while it was eating.

Teratorns were very large birds. One teratorn fossil found at Rancho La Brea was a Merriam's teratorn. This incredible bird stood nearly 80 centimeters (31 in.) high and weighed more than 13 kilograms (30 lb). The distance from the tip of one wing to the tip of the other wing was about 3.3 meters (11 ft)! Scientists think that teratorns ate small animals such as frogs, lizards, mice, other small birds, and fish.

Dire Wolves

Probably the best known fossils unearthed at Rancho La Brea are those of mammals. Mammals are warm-blooded vertebrates. They have hair or fur that covers much of their bodies. Mammals care for their young and produce milk to feed them.

The most common large mammal fossils pulled from the pits at Rancho La Brea are those of the dire wolf, which is now extinct. The bones of several thousand dire wolves have been collected. Like today's wolves, dire wolves probably lived in packs, so many of them may have become stuck at the same time.

Dire wolves had very large teeth and probably ate mostly moose, deer, bison, horses, and other large mammals.

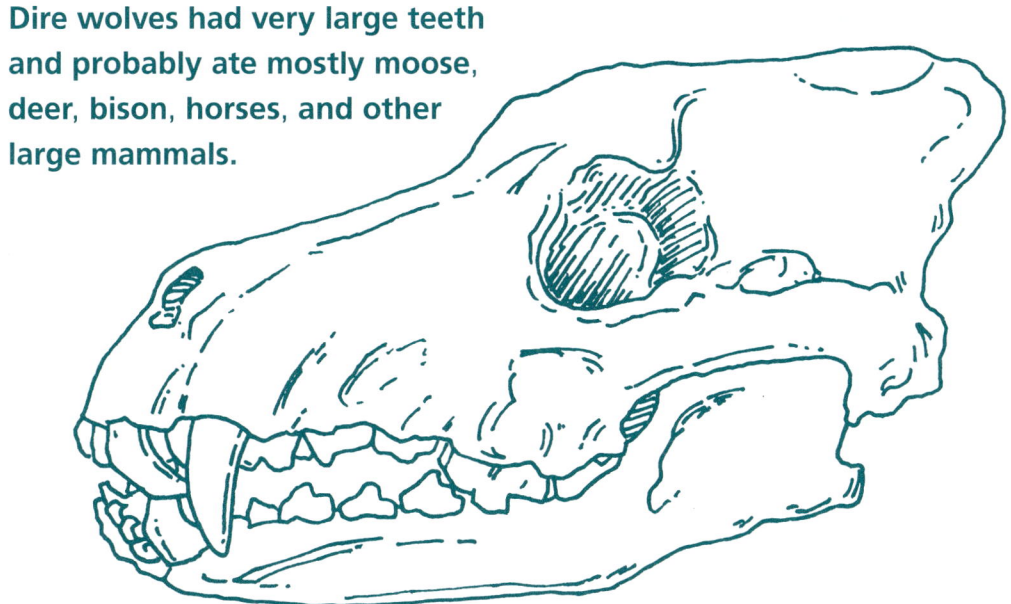

Unlike their modern relatives, dire wolves had short, sturdy limbs. Dire wolves were about 1.5 meters (5 ft) long and weighed about 50 kilograms (110 lb). These wolves had bigger teeth than do the wolves that live in the western United States today. Dire wolves probably hunted for much of their food. They also ate the bones and flesh of dead animals.

Although dire wolves had fairly large teeth, their skulls were much smaller than the skulls of modern wolves. Even though the brains of the dire wolves rotted away in the asphalt, scientists are able to tell that these brains were much smaller than the brains of modern wolves.

Saber-Toothed Cats

The fossils of many saber-toothed cats have been discovered at Rancho La Brea. In fact, the bones of about 2000 cats have been found. Like dire wolves, saber-toothed cats are extinct. Scientists believe that the last of these mammals died about 10,000 years ago.

The saber-toothed cats preserved at Rancho La Brea are about the size of a modern African lion. However, saber-toothed cats were probably about twice as heavy as their modern relatives. Saber-toothed cats were very muscular and had very strong limbs. They also had short tails, which means that they probably didn't run for long distances to capture their prey.

Instead, the saber-toothed cats probably used their powerful limbs to move quietly through grasses to capture prey. After capturing an animal, the saber-toothed cat tore its prey apart with its enormous front teeth—which were nearly 18 centimeters (7 in.) long! Ancient horses, buffalo, deer, and antelope were just some of the unfortunate prey of the saber-toothed cats. These cats probably also ate animals that had been killed by other large mammals.

The pattern of a saber-toothed cat's fur probably depended on where it lived. A cat that lived in an area with few trees might have had a spotted coat, or a coat of only one color. A cat that lived in a dense forest might have had a striped coat.

Like present-day lions, saber-toothed cats probably lived in groups. Fossils show that the youngest cats—the kittens—were cared for by the female cats for a while after birth. Scientists have also found bones from saber-toothed cats that show that the bone had been injured but had healed. This indicates that saber-toothed cats may have cared for injured members of a group until their wounds healed.

Although they were caring, saber-toothed cats could be very aggressive. Their fossil throat bones indicate that these cats were able to roar to chase away unwanted visitors. Some of the bones found have holes in them as large as the teeth of these cats. This shows that they sometimes may have attacked one another. Like today's lions, saber-toothed cats may have fought over food or hunting territory.

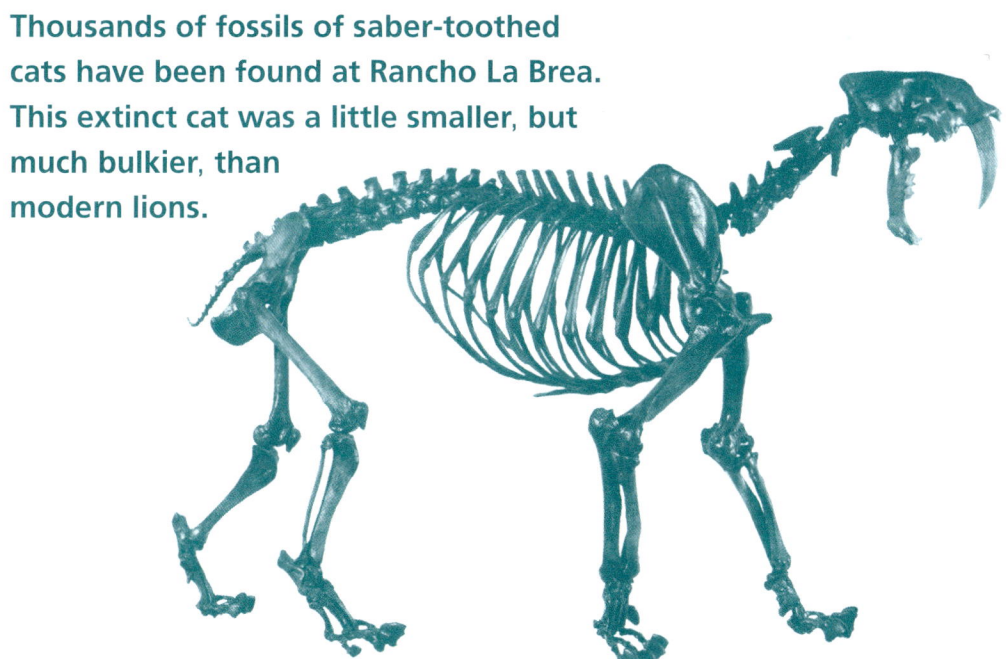

Thousands of fossils of saber-toothed cats have been found at Rancho La Brea. This extinct cat was a little smaller, but much bulkier, than modern lions.

Mammoths and Mastodons

Mammoths and mastodons are the largest animals preserved in the asphalt at Rancho La Brea. These two species of mammals are closely related, but they are very different, too.

The mammoths trapped at Rancho La Brea were Columbian mammoths. The mastodons were American mastodons. Columbian mammoths were much larger than their mastodon relatives. From its shoulder to the ground, a typical Rancho La Brea mammoth measured just a little less than 3.7 meters (12 ft)! One of these mammals could weigh as much as 9 metric tons (10 tons). The American mastodon only measured between about 2 and 3 meters (7 and 10 ft) from its shoulder to the ground. A mature American mastodon probably weighed about 5.4 metric tons (6 tons).

Mammoths and mastodons were the largest creatures preserved at Rancho La Brea. The mastodons were much smaller than their relatives, the mammoths.

Rancho La Brea mammoths had four broad flat teeth that were adapted for chewing grasses. Mammoths continually grew new teeth to replace worn teeth, and they had a total of six sets of teeth in a lifetime. Rancho La Brea mastodons had teeth that were adapted to chewing leaves and twigs. They looked a bit like human molars.

Both the mammoths and the mastodons had tusks. The mammoths' tusks were longer and much more curved than the mastodons' tusks. Both animals used their tusks to defend themselves and to help gather food. The tusks of the mastodons were much shorter than those of the mammoths. The mastodons' tusks were almost parallel to each other and curved upward slightly at the ends. Some studies have found that the inside of one of a pair of tusks is usually more worn than the other tusk. This might mean that these large mammals favored one tusk over the other.

Both the Columbian mammoth and the American mastodon are extinct. No one knows for sure why these large creatures died out. One hypothesis is that a change in climate may have reduced their food supplies. Another possible reason may have been disease. A third possible reason for their disappearance is overhunting by humans.

Bears and Bison

The short-faced bear was the largest carnivore preserved at Rancho La Brea. The animal's large, sharp teeth indicate that it ate much meat to supply energy to its 800-kilogram (1800-lb) body. Standing on its hind legs, this bear was nearly 3.4 meters (11 ft) tall! The short-faced bear's huge nasal passage and eyes that faced to the front suggest that the animal had excellent senses of smell and sight. This extinct bear had very long legs, and its toes were not turned in. These adaptations allowed it to run quickly after its prey.

The short-faced bear was larger than today's grizzly bears and polar bears.

The ancient bison, a relative of modern North American bison, is the most common mammalian herbivore found at Rancho La Brea. The large heads of these animals were adapted to push away snow to get to the grass below the snow. The hair of these bison prevented the animals from losing body heat. Like the bison that live today, the ancient bison lived and traveled in herds. These animals migrated and spent only a few months each year at Rancho La Brea. Fossil evidence suggests that these mammals were near the asphalt pools only during the spring.

La Brea is Still Catching

Oil is still seeping to the surface at Rancho La Brea. About 32 to 48 liters (8–12 gallons) of new oil surfaces each day, and animals still get trapped in it, especially when it is warm. In 2003, a flock of 60 cedar waxwing songbirds got stuck in the goo.